BLOOD ON THE COAL

BLOOD ON THE COAL

The Story of the Springhill Mining Disasters

by

Roger David Brown

LANCELOT PRESS

Windsor, Nova Scotia

DEDICATION

To
The people of Springhill
and
My good friend Blair Kindervater

Published 1976

LANCELOT PRESS LIMITED, Windsor, N. S.

ISBN 0 — 88999 — 048 — 4

PROLOGUE

SPRINGHILL

This town has a story to tell — of tragedy, of valour and of tenacity. This is a town that refused to die. Here, at least 424 people have perished mining for coal.

1891 — it opened as a good year for Springhill. Its population was nearing five thousand; a bright future was assured by its location on top of a great underground hoard of the black gold of the nineteenth century, coal.

A mere 31 years before, the Springhill region contained about five small houses and perhaps twenty people. This was the meagre population after all the years of settlement since 1790. For it had been in the spring of 1790 that three young United Empire Loyalists who had served with George III's army at Harper's Ferry, Virginia, left the refugee loyalist settlement at Sackville,New Brunswick, and claimed their land grants in what is now Springhill. These three were John and Charles Boss and David Herrett. Later a few others came, including John and William Hunter who settled on the "Hill" and likely gave it the name Spring Hill because of the abundance of springs in the area.

By 1827 local people had found coal. It is known that Lodewick Hunter, in 1834, was selling coal to nearby blacksmiths. On April 26, 1849, the General Mining Association of London, on a lease from the Duke of York, who held the Royal Grant for mines and minerals, secured 160 acres of land east of Lodewick Hunter's grant. Soon it got another 2000 acres. This company operated mines in Cape Breton and Pictou. In 1858 it lost its monopoly.

Others took up the challenge, the challenge of opening a real mine in Springhill.

On April 18, 1870, the Springhill Mining Company was incorporated by law. Soon a rail link had been laid to join the national track system at Springhill Junction, and another line had been driven nearly thirty miles to the Bay of Fundy port of Parrsboro. The 1880's saw a new company take over, the Cumberland Railway and Coal Company. Employment in the three mines had reached 1,350 males and 60-70 horses. Springhill was on the map. Yes, 1891 looked like it would be a good year. It wasn't.

1891 — It's hard to imagine now that by-gone age. There weren't even any deer in Nova Scotia until 1894. My grandfather tells of the first deer he ever saw, "I shot the animal and I didn't know what it was. I went home and told my father and he ran up into the woods. 'It's a deer,' he says."

Sad to say he also tells of his Uncle Jim and Uncle Albert Brown — they shot a buck and a doe caribou — the last of these animals he remembers hearing about around here. The time was around the turn of the century.

To make this account as realistic as possible I have tried to use the miners' own words as they told their stories and have tried to make it as accurate as possible.

CONTENTS

MAYOR GILROY

CHAPTER 1

"Her Majesty has heard with much regret"

THE EXPLOSION OF 1891

Saturday, February 21, 1891: That morning Charles McNutt entered Jewkes Brothers' store to buy tobacco. He cut his tobacco with his knife and then left to go into the mines. Behind him, in the store, he had forgotten his knife. He never came back. His name can be found in the list at the back of this book; his knife can be seen today in the Springhill museum.

It was seventeen minutes to one. Fourteen year old Danny Robertson, feeling refreshed after his half-hour dinner break, was back at work driving his rake of empty coal cars. He was a good looking boy with dark hair and a square jaw. At this moment Jenny, his horse, was pulling the rake into the level on the west side of the mine.

Without warning, a shot of fiery air hit him, knocking him backwards into an empty coal car. For many seconds, perhaps minutes, he lay there dazed - the thunder of falling rocks and timbers echoing in his ears. As he got up from the box-car he became aware that his clothes were burning. Jenny - Jenny was dead. And, except for the bits of flame, all was darkness. He found, as he threw off his burning coat and vest, that his hands and arms were badly burned. Fighting his fear, he began to attempt to find his way from the

underground tomb.

A boyish cry for help caught his attention. Making his way to these sounds of hopeless panic, feeling around in the bitter blackness, he found little Willie Terris, the trapper. Willie was twelve years old. His job in the mines was to open and close doors in the passageways through which men and coal moved. These doors, as well, controlled the air currents which were used for ventilation. Willie was petrified with terror; but he was alive in the bowels of the earth, and except for small burns on his hands and ears he was all right. He had been sitting by the door and when he saw the flash of flame rushing toward him, he had dodged under his seat, burying his face in his hands.

Robertson, himself, was too badly burned to take a grip on the younger boy and lead him to safety. So he carried him piggy back style, on his back, through the mine to a place of safety. Robertson asked the rescue party where his brother was. They didn't know. He wanted to reenter the level and search for him. But the rescuers sent him in a box with some other wounded to the surface. And there was his brother.

Morrow, the author of a book on Springhill written shortly after the explosion, relates the story of another boy: thirteen year old John Conway. His father was the underground manager of the number one slope. Like Robertson, he was driving a rake, sitting on the front box. The explosion killed his horse, pinning him, uninjured, under the animal. There he lay for an hour until a discouraged search party heard a weak voice calling for his mother. Listening to the cry, "Mother," they found the horse and the boy.

Ironically John's father had checked out the number one mine slope that morning and found everything "in perfect condition, and the air circulating freely."

What had caused the explosion? The following section is reconstructed from the inquiry into the explosion held in the spring of 1891.

"A person could get lost in the mines," the former

miner told me. Underground, from the main shaft, at certain depths in the ground, long side passageways called levels were driven and from these came a great maze of tunnels.

It was in one of the smaller tunnels, the number three bord in fact, that Thomas Wilson was working at the time of the blast. Underground manager Conway later stated that the morning of the blast Wilson had told him that "everything was perfectly clear and in good condition." It was Wilson's job to blow coal from the face of the seam. This was done by boring a small round hole into the wall, placing an explosive in this hole and then packing it to keep the flame from shooting out into the mine.

At 12:43 p.m. he set off his charge and a terrific explosion thrust out from the number 3 bord of number 7 balance in the west side of the number one mine. Even the jury at the inquiry failed to be certain as to what had gone wrong. Searchers found Wilson and two others in the bord. It appeared they had been walking from here when the trouble broke behind them. Wilson was lying across the bodies of the other two - all three had their faces to the ground; all three were not disfigured. The jury felt there had been "an unusual slip in the stone." In other words, a crack in the rock had allowed some of the flame to shoot into the mine "igniting the coal dust and a certain portion of the gas which might have been present at the time." Because of this accident one hundred and twenty-five men and boys died.

As to what happened next we have the testimony of a man who had spent fourteen years in the mines at Springhill, Patrick Hennesey. The flames shot through the mine, drying the watered down walls, and feeding on the dust and gas liberated by its drying affect and the chaos of falling timbers. Hennesey said that a terrific gust of wind and coal flew up the mine. Flinging himself on the floor, he saw the flame sweep by him with a roar. Five feet away he saw Henry Nash being burned seriously. Then the flames came back as the force retracted toward its origin, leaving behind death and destruction. The air was suffocatingly hot.

The main cause of death in the number one (East) mine was the explosive force while in the connected number two (West) mine it was the dreaded poisonous gas.

Gas is always present in a mine, a product of the decomposition of the plants that formed most of the coal. This fire-damp gas, when mixed with the right amount of air, is extremely explosive. Fortunately it is lighter than air and will rise out of the mine. One of the mining dangers is that it can be present in a hollow chamber in the coal bed and can burst out when the miner's drill reaches this cavity.

After fire-damp burns, it leaves behind poisonous gases such as after-damp and black-damp, both heavier than air, and the most dangerous of all, white-damp: a gas without colour, taste, or odour, a gas lighter than air. A breath of it spells death.

The explosion at Springhill was different from the two other major mine disasters the town was to endure. After two o'clock - that is a little over an hour after the explosion - no survivors came out of the mine: there was no miracle. There were instead 121 dead, a mere 17 injured, four of whom were to die of their injuries. The number of dead included 17 boys sixteen years of age and under.

Another unusual aspect was that there was so little physical damage done to the mine. True, near the explosion site were scenes of virtually complete destruction, but it was only a few days after all the dead were removed that the miners returned to work as usual.

The *Halifax Morning Herald* of Wednesday, March 4, carried this ad:

Wanted At Once
One Hundred Miners
To Work In The
Spring Hill Colleries

DANNIE ROBERTSON.

"Danny Robertson!" said Lance Swift, "He was the hero of the 1891 blast. He died out west a few years ago. Edmonton, I think."

Ninety-five year old Bill Graven added: "Willie Terris, the boy that was rescued by Robertson, Terris, he never went back into the mines."

Earlier the men had refused to reenter the mines to work until all the dead had been removed.

Morrow relates in his 1891 book the first hand statements of the Rev. David Wright, the Presbyterian minister of Springhill. He entered the mine on Saturday shortly after the explosion:

"The scene along the level was pitiful. The explosion had rushed toward the bottom of the mine. Not much damage was done, only the props knocked out, which caused the sides and roof partially to fall in. The first thing we saw on our way in, after leaving the bottom of the slope, was a couple of horses lying dead. A little further on and two others were standing, one of them literally burned red - at least the side next to us was red flesh. It was killed. Then right in from these the debris blocked the way. Now and again a large fall, under which was a horse, part of it visible. The driver was, we knew, somewhere near."

He went down again Sunday. Since he was holding the lamp, he was at the lead of a small party carrying a body.

"When we reached number four balance the roof was falling in. I was in front and drew back. James Harvey went on to examine, and in a few minutes reported that it was not safe to go on, and we sat down for 15 or 20 minutes. We were here joined by four or five others carrying another dead body. After the falling had ceased, Harvey examined again, and reported all safe, and we pushed onward and reached the bottom of the slope in safety. It was a weird sight in one place to see two bodies sitting each with a safety lamp in his hand."

News of the disaster was flashed around the world. Halifax newspapers carried reports from their on-the-spot reporters on "the Springhill Calamity". Their facts are sometimes inaccurate and their language sounds odd today. The *Halifax Morning Herald* , for example, describes the death of the little Dupee boy who ran to see the corpse of his brother after it had been brought from the mines. He "fell upon the frosted ground and struck his forehead with such

HENRY SWIFT,
Late Underground Manager Springhill Collieries.

"Well, my name is Lancelot Swift and Henry Swift, the underground manager, was my father," said the old gentleman gazing out the sun-porch window. "He was down there six dys, you know - the last one they brought up.

"He was down there late that day. There had been a story going around that an old Pictou woman, Mrs. Coo, had predicted trouble in the mines. So, when my father was ready to come up, he said to a miner down there, 'I've checked it all out and I can't find anything wrong but I'm going to check again.' They found him buried under rubble.

"Mother had to move out of the underground manager's house, "he said, pointing to the vicinity of Brown's funeral home, "She bought this one here in 1892 - been here ever since. I'm sorry I can't help you more but I don't remember much about my father because I was only three when he died."

violence that he knocked his brains out. He never moved after his fall." He was four years old.

The *Halifax Morning Chronicle* under the dateline of Thursday, February 26, 1891, conveys to its readers a picture of Henry Nash, the man who had been badly burned as he lay near Hennesey:

"wounded Henry Nash died at midnight. He was terribly burned about the face and hands, he never rallied from the shock. After...(the)...explosion...(he was) very nervous...(it was) intensified owing to Nash living on the street leading to the cemetery and insisted that he should be allowed to see the funeral passing. Dr. Cove ordered the blinds to be drawn down, but Nash induced his friends to disobey the doctor's instructions. He was completely prostrated and sank rapidly...

"...became raving maniac. It took three men to hold him. He cried upon those in the room to save him and then madly raved and wanted to be allowed to go into the pit to save his comrads...exhausted he died."

From Springhill's Mayor Hall came this appeal to be published in world newspapers:

"to the public: A mining disaster, attended with fatal results unparalled in the history of Canadian mines has fallen upon the town and the people of Springhill...the widow and fatherless will require abundant assistance and that promptly from a public shocked and horrified by their horrible calamity...(we) ask for immediate response of all denominations, societies, guild trades and nationalities."

Money and aid, far more than was asked for, came pouring in. In two months the $70,000 goal was reached. In a little over three months the city and populace of Montreal had given $22,000 dollars, followed by $16,000 from Halifax; $8,000 from Saint John; $3,000 from London, England; down to small donations from villages such as $40.40 from Upper Musquodoboit, Nova Scotia.

Telegrams of concern came from Queen Victoria,

who cabled through the Governor-General, Lord Stanley:

> Her Majesty commands to inform you that Her Majesty has heard with much regret of the Springhill colliery disaster, and desires me to convey an expression of her sympathy with the injured and with the relatives of those who have lost their lives. Be good enough to send details for the Queen's information.

From Springhill came an answering telegram:

> Springhill Mines, N.S., Feb.23
>
> Lord Stanley of Preston, Ottawa:
> Her Majesty's devoted subjects are deeply grateful for Her gracious expression of sympathy in their sore affliction.
> Please transmit to the Queen our warmest thanks.
> Henry Nash, one of the wounded, died last night. The recovery of others is doubtful.
>
> William Hall, Mayor

Lord Stanley, the man who gave Canadian hockey the Stanley Cup, donated $500 dollars to the disaster relief. The Queen also gave a donation.

Springhill recovered. In 1892, for example, its electric light system was installed. But as the years rolled by there were always fatal accidents in the mines: 182 men died in well over 125 separate accidents between the '92 and '56 explosions. On the negative side as well there was a bitter 22 month coal mine strike from 1909-11. And then there were the two world wars with a high rate of enlistment in Springhill.

By 1956 the population stood at nearly eight thousand with 1,585 employed in the two mines, 1,200 of these underground.

Queen Elizabeth II and Prince Philip
in Springhill Junction, 1954.

Ernest Brown and
son Carl, in 1958.

"Well, I was a scab," said 86 year old Ernest Brown of Southampton, "or that's what they called us. It was a twenty-two month strike, you know. I started in 1909, before I was married. The strikers came out here to work; they took our jobs in the forests....They'd work cheaper than we would. They were drawing money out of the union, too. They could afford to work cheap.

"At the mine property entrance there was a group of strikers and the police would part them to let us through. They put me to work fixing company houses. A band of strikers came out and drove Henry Brown, my uncle, and me from the roof with stones. After that they put me to work in the carpenter shop. Coal boxes used to come up with broken boards in them and we repaired them.

"My father used to drive us into town from the country in a wagon after the weekend. Then he got scared and stopped. The company gave us a hand-car and eight or nine of us came out by South Brook this way on the rails. I got off there and the car went down the track to Parrsboro. Then they stopped us."

"How did they stop you?" I asked.

"They threw stones at us. They'd corner people and beat them, they would. Then the company gave us a pass to go on the train and we'd knock off work on Saturday at noon. I was married by this time. Then Hubert Skidmore and I got the measles from the people we boarded with and that finished it. I didn't go back to work for the Company until the strike was over."

Then he went on to describe the mines as he saw them in 1913. "A miserable place. We'd go down on the rake there - eighty people at a time. Oh gosh, I dreaded going down on that thing. Before we'd go down we would go to Bill Graven in the lamp cabin, give him the metal tag and get a lamp. Then someone would search our pockets to make sure we carried no matches into the mine. A person could chew tobacco down there, though. If you wanted to get along with the young fellows down there - give them a chew of tabacco. Worse time I was sick was in there - had to come home, swallowed some tobacco, I guess. But getting back to the lamps. In the mine at the bottom a man would blow in the lamp to see if it were safe. Go out awful easy, they would. Then a person would have to go out to the level to a place that was all bricked in and they'd light it for you.

"Yes, I worked in the old number two at the 6500 foot level. I loaded coal in box cars. Got $1.52 a day, and had to board myself. Some of them called me a scab - if they thought you'd worked in the mine during the strike. They called me buckwheat eater because I was from the country.

"I worked there three months. They still used horses down in the mines. They'd haul the cars on underground rail-track. I quit. Why? I didn't like the job. All that black coal dust and dirt, and it was dangerous."

And this is how a country person viewed life in the coal mines. He went back to the forest.

CHAPTER 2

Faith, Hope, and Charity

THE EXPLOSION OF 1956

As November, 1956, began, provincial newspapers were filled with accounts of the Russian invasion of Hungary and the Anglo-French battle to regain the Suez Canal in Egypt. In addition the province had a new premier, a Conservative. Robert L. Stanfield had just two days before ended 33 years of Liberal power in Nova Scotia and had begun a career which would see him go on to become the National Conservative leader and opponent of Prime Minister Trudeau.

But in Springhill what thoughts would have raced through the minds of any miners if they had foreseen the loaded coal cars tumbling backwards down the auxiliary slope that Thursday afternoon.

It wasn't the first time such an accident had happened in the number four mine. On different occasions in 1955 and 1956 cars had become uncoupled and run amok in the mine. The Royal Commission which investigated the 1956 blast reported that on January 11, 1956, an empty train going down in the mine had derailed at the 4800 foot level (1464 metres). It had cut the armoured electric power cable and a fire had broken out in the roof supports. Now, eleven months later, no action had been taken to prevent a recurrence of a similar accident. The general opinion seemed to be that the grade was too steep to be able to eliminate this hazard.

The number four mine was, as mines go, a fairly new mine, having been opened in 1936. It had reached a vertical depth of 1063 metres or 3486 feet. On this November 1 afternoon there were 118 men underground and 42 on the surface in the second shift of the day. It was to be the final shift for this mine.

The six cars crashing down the steep 35 degree slope were from a seven car train which, while being hoisted to the surface, had become unhooked. Here and there they smashed into an armoured steel transmission cable which was carrying 2200 volts. Coal dust rose in billows in the air. And then, nine metres above the 4400 foot level the cable was crushed. The time was 5:07 p.m. Flames shot out as a violent explosion ripped through the mine. Adding to the force of the blast was coal dust blown from another train which was being brought to the surface.

The bolt of fire shot to the surface. The Bankhead building which covered the entrance to mine number four crumbled in the fiery blast, only one wall surviving the force. In these ruins five men died immediately; two others were to perish from their burns. Two men here, Norman Boss and Alan Skidmore, had miraculous escapes. These two were blown through the door onto the ground and although they were burned and suffered from shock, they survived.

In the town all life stopped. The bluish flame rose sixty metres (200 feet) into the air and disappeared. The sound echoed over the hill. Then came the wail of the siren; the rush toward the mine.

Speedily the nearby number two mine was shut down. The local fire department arrived on the scene to help fight the flames at the Bankhead.

Draegermen are the rescue workers of the mine, or more correctly the trained rescue workers. They owe their name to a German, Alexander Draeger, who devised special equipment for breathing in a gas filled mine. Within a short time local men formed their draeger crews and began to penetrate the mine. The *Springhill Record's* special disaster edition of 1959 carries these words:

> When only about 100 feet down from the mouth of the steep, debris-littered slope, several of them were overcome and it was only with the greatest difficulty that their mates were able to drag them out. This activity was greatly dramatized by the fact that men standing at the mouth could clearly see, by the light of the draegermen's lamps, the desperate struggle going on; yet they could do nothing to assist the rescuemen because of the presence of the invisible but deadly gas. One of the men stationed at the mouth of the slope said that the concentration of gas was so heavy that it extinguished a safety lamp only a few feet down.

In these initial probes two draegermen lost their lives.

When the dawn broke on Friday the death toll stood at thirteen. Due to heavy smoke the order was given that day to seal the mine for two or three hours as a safeguard against a further explosion. At six that evening the mine was reopened.

On that sad Friday morning Dosco General Manager Harold Gordon had told a group of forty newspapermen, "There's just the slightest possibility that some of them are alive. The probability is that all are dead." Mine Manager George Calder, as well, held out little hope.

Most of the men who had been working in the mine at the time of the blast were unaware that an explosion had taken place, as Harold Tabor tells in the *Halifax Chronicle-Herald* of November 5, 1956:

> I was operating a cutting machine at the 5,700 foot foot level when it happened. At that depth we hardly felt a thing. George Stonehouse (since saved) came tearing up the wall and said the power was cut off and to get out fast.

We started making our way through the turn to the slope when the smell of smoke hit. It smelled like old rubber and everything burned and choked us. We thought the rake had jumped and we knew that a fire had started, but we never thought that there had been an explosion.

We tried to get to the slope - Donald Campbell, Joe Holloway, Don Hopkins, and myself - but men were dropping like flies everywhere. Everybody was saying help this fellow and that fellow. We came across Victor Henwood, who had collapsed. We grabbed him and dragged him to where the smoke had stopped.

Then all four of us collapsed with him.

The gas was getting stronger. All in all I fell to the gas fumes five times during the time we were down there. I shook some water over my coat and held it to my face. The rest did the same and we made our way back to the 5700 foot wall from the slope - the place we were in the first place. The air was fairly good here, but it was getting deader all the time. We made our way to the 5400 foot level, where we turned on every air valve that we could. The younger fellows were all collapsing.

There was a bit of panic atmosphere by this time. What were we to do? Some of the men pulled out old stoppings and tried to get out that way. They never came back.

At this point there were 47 seven of us. Con Embree said we should go out to the "turn". Now there was a man! If ever there was a hero it's him. He had been through two previous explosions and we knew anyone in trouble with Con Embree was going to come out of it. So Con and I set out. This was Friday morning. We knew there was a phone in the "turn". The wires were broken, so we spliced them, but the line was dead.

We could see the smoke coming down at us. We went back to the men but we didn't dare tell them about the onrushing smoke and gas. We made a makeshift door out of old trapdoors and bratish cloth and diverted the old dead gas to an old tunnel. Now we had the gas boxed off. We knew no gas could get in and we had compressed air. We knew as long as they kept the pressure on we could live.

It was Deputy Overman Conrad Embree who became the leader of the trapped men. In the diary he kept he recorded the underground drama. His father had been one of those who survived the 1891 blast. He, himself, had started mining when he was fifteen years old. In the 1920's he had survived two mine explosions - one in the Chignecto mine and the other in the River Hebert mine.

Here, in the shelter the men talked of home and religion and the fact that it was Friday - pay day. At different times a few men would leave the shelter to see if any other survivors had staggered toward it. To help filter out the gas they ripped chunks of cloth from their clothing, soaked them in the dirty water which had gathered on the floor and held the fabrics to their mouths. These "filters" enabled them to stay "outside" for only a few minutes.

One group of six men was found by one of these scouting parties. To Emile Melanson, one of these scouts, it was a joyous moment - one of the six was his brother, Leon.

Tabor continues, "Joe MacDonald, one of the underground mine officials, and a party of four men made another trip out to the slope. It was about 300 to 500 feet to the hoisting slope.

"Three other men were found out there, but only Doug Beaton of the three was alive. We found Dougie with part of his face in a stream of water that was flowing down the slope. This was why he was alive. He apparently got enough oxygen from the water.

"Under the direction of Frank Hunter, a former draegerman, we took turns applying artificial respiration, and kept it up for five hours before Burton was revived. We placed him in an old plaster box and covered him stone dust bags. There were no blankets or coats. We had abandoned them getting to the level. We placed him by an air valve. This about did us in."

Embree, himself, wrote of the rescue of Beaton, "Got out to the 5400 bottom; after-damp bad, but seemed to be improving some. Found Doug Beaton about 25 feet below the 5400 back slope bottom and got him inside the door. Very little sign of life; applied artificial respiration. He showed some signs of recovery by 3:15 a.m. with men taking turns keeping up artificial respiration for about four hours."

Outside the shelter signs were chalked up for the rescue crew:

FOR GOD'S SAKE COME AS FAST AS YOU CAN

and

47 MEN ALIVE HERE

"Then," adds Tabor, "the water really started to come down and the after-damp (gas) started to be more evident. We went back to the temporary place we had made and stayed there until 2 a.m. Saturday. From here various ones would from time to time go out to see if the air in the turn and the slope was improving. We made a sign and every time someone went out, we would mark it down.

"It's still hanging down there. Avard Glennie, my buddy, went up the back slope and pounded on the pipes. He gave eight raps. We learned later the men on the 4100 level also rapped eight times but because of a broken pipe we couldn't hear them and they couldn't hear us. Then Avard Glennie and Harold Lewis decided to try to make their way out. They went out and we never saw them again. I think I'd know where to find them.

"We made another trip out to the slope and found the gas to be about 7 per cent, enough to kill a horse. At the same time the water stopped coming. We figured they had sealed us off and we were going to die like rats. After that the after-damp got worse. Conn Embree then showed the way that saved our lives."

Embree's solution was for some of the men to go with him and get some two and one half inch rubber hose which had been used in the mine at the bottom to convey compressed air to the machinery. When they brought the air hose back they cut holes, about one foot apart, in it, on opposite sides.

Tabor adds, "Then we plugged our noses with pieces of cloth and breathed from the air hose cuts through our mouths. It was now Saturday. And we knew we had to do something.

"Charlie Burton had a red safety light. Six of us decided to make a try for it through the fanway. They were Charlie Burton, Lorne Dakin, Murray Terris, George Stonehouse, Lewis Austin, Ron Stonehouse and myself."

They had a distance of 670 metres (2200 feet) to go. The track through the fanway meant a steep climb ending in 120 metres straight up. Embree had written in his diary "the men are growing weaker." He said later, after the rescue, of those who left the shelter that morning, "I figured they had one chance in a million."

"I knew," Tabor relates, "there was one air value at the 5100 foot level, and when we got there, we turned it on, got filled up with fresh air and took off again. The next air valve was at the 4800 level.

"On the way, George Stonehouse made breaks in the joints of the air hose to let more air out. We made only small breaks, to make sure there still would be a flow of air to the men we had left below. We got to the 4800 level and got another breather there from the air valve. The next level was 4600, but there was no air valve there....The gas nearly got us on this jaunt. And when we finally did get there...(the next air valve) we were surprised to find camped there 23 other men who had been working...on the...wall. These men had an air compressor in operation."

Where had these men that Tabor reached come from? They had been working deep in the mine. Wilfred Brown, one of the leaders of this group had been at the 6100 foot level. At 5:07 p.m. on that fateful Thursday as Deputy Overman Brown was bent measuring the depth of a newly bored drill hole, he heard the omnious sound. There were 57 men working on this, the deepest level. "Knock off work," were the orders. Then began the retreat from the mine. Different groups chose different exits. Twenty-five men followed Brown to the 5700 foot level. Here they managed to drag some miners from a gas filled pocket, but the fumes were too strong - some men they couldn't reach.

Luckily they found a way up to Con Embree's shelter at the 5400 foot level. After a few minutes in this shelter Brown and some of his group decided to try to get out of the mine. However, most of this group decided to return to Embree's shelter as Wilfred Brown was having difficulty finding his way.

Six men - Peter MacDonald, Wayne Smith, Sonny Michaelson, Kenny Gilbert, Donald Campbell and Dick Ward - stayed with him. For much of their trip the going was easy but at the 4400 foot level Brown's safety lamp went out, they were in gas again. Luck was with them once more. They were able to build a shelter around an air valve and here they stayed. "Fellows," said Brown, "it's my birthday." The time was in the early hours of Friday morning. And here in this small enclosure the time old song rang out, "Happy birthday to you, happy birthday to you; happy birthday, dear Wilfred, happy birthday to you."

As time went by another group joined the seven until their number had grown to twenty-eight when Tabor arrived.

Encouraged, they set off, as Tabor relates; "We then started up. We couldn't punch holes along the way because the air hose at this level was too big. We reached the 4100 level where we were able to get a new supply of air from a valve. Now we had to make our most desperate plunge to get up further. We had to go up an almost perpendicular ladder, a distance of between 300 to 400 feet, to the 3200 level. This would get us to the top of the back slide. We soaked our shirts in stagnant water from the electric coolers, put them over our faces to keep from breathing the gas, and started up. We passed the water can up and down to each other as the cloths over our faces dried. And they would become bone dry with about every ten steps up.

"That was how bad it was there. By this time some of the 23 men we had come across joined us. We made the 3800 foot level. George Stonehouse and myself were the last two in this trek. I was behind George and I could hear him moan and groan. When I got up to George we both started to pray and I don't care who knows it, we prayed. If only we could make it to the 3500 level, we felt someone would get out.

"George made it but I flaked out. My light was still on. I remember every time I stopped I had turned off my light, so had saved it.

"I recovered enough to start to crawl, and then I heard the voice of Lewis Frost, the chief mining engineer of the mine, and a draeger crew. They had a stretcher and put the oxygen to me. Then they took me from the 3500 level to the 3200 level. They carried me while making their way on their hands and knees. With all that equipment they did not have much room. If they even raised their backs they would strike the roof top of the tunnel. At the 3200 level, I got my first big gulp of fresh air. I was asked if I wanted some coffee or a drink of brandy or rum. I took the rum and they filled one of those dixie cups full. Did I want some ginger ale or a mix? I said no and downed it all."

That Saturday, just before noon, the town was electrified that there were survivors. The first up was Charlie Burton.

For the men in Embree's shelter it was still to be a long wait. Unknown to them gas leaks in the mine were making it slow progress for the rescue workers, both the draegermen and the large number of bare faced volunteer men. Many of these workers had to be brought back to the surface because they became sick. But they returned again to the mines.

Embree on Sunday evening held up his diary to the men. He planned to write no more he said. With the pages left they could write wills. Several men wrote in the book. Despite a depressed feeling the men joined Archie Legere in singing one of the big hits of the '50's:

Have Faith, Hope and Charity
That's the way to live successfully
How do I know?
The Bible tells me so.

Don't worry about tomorrow
Just be real good today;
The Lord is right beside you:
He'll guide you all the way.

Then later that evening the town was overjoyed as the draegermen sent word to the surface that they wanted some raisins for more survivors.

"There are at least 50 alive," people told each other.

Embree described the rescue this way: "We heard someone banging at the door. When we opened it, they were there, led by McCormack. Everyone wanted to hug him," "In a few hours the 52 men from Embree's shelter were brought to the surface. As the leader was brought to the surface, he insisted on walking out to the ambulance. "I want to go out of here the same way I entered," he said. At the hospital Embree was carried in by the miners themselves - a very personal thank you.

"What was the big thing that stands out in your mind about it all?" he was asked.

"The men. They were really wonderful. There wasn't one bit of panic. Yes we sang. We sang a lot. We had just finished a duet when the draegermen broke through to rescue us."

And the rescue operation was over by 6 o'clock Monday morning. There had been 88 survivors underground, some having been trapped well over 80 hours. And there had been 39 deaths - 30 of them underground. Even 19 year old Philip Brown who had been tending a pump at the 8400 foot (2360 metres) level had survived. He had been the person deepest in the mine.

And at 7 Monday morning the explosion doors leading from number two mine to the number four mine were closed. Gas tests had shown that no other men could possibly be alive in the mine. There was danger of a second explosion and so it was agreed to seal the mine. The 26 bodies underground would be recovered when it was felt to be safe to reenter the mine. By Wednesday afternoon the mine was completely sealed.

It remained closed until January 18, 1957 when it was reopened to remove the bodies. And that was the end of mine number four.

CHAPTER 3

Another Thursday

THE FIRE OF 1957

The *Halifax Chronicle-Herald* in mid February, 1957, recorded the views of the townfolk on the future of Springhill. "Everybody has been hit hard in some way or another," said Sandy Leadbetter who ran a men's clothing store on main street. "It'll be a year before merchants find things back to normal." "Flames would destroy his shop in less than a year.

"With a lot of unemployment you'd expect trouble, but there's been none," said Police Chief Leo MacDonald.

"When the number four pit re-opens," people said, "it'll be the safest mine in Canada." It would never re-open.

However Mayor Gilroy, looking into the future, said "...we've had to postpone plans for a junior and a senior high school, but we haven't cancelled them and that's the main thing." Today this school has been built.

There was money to help the people: the disaster relief fund for the 1956 blast would reach one half million dollars.

Main street in Springhill before the fire reminded me, as a child, of a western main street as portrayed on television - false fronts on wooden buildings. They burned well.

Christmas 1957 passed. Over a year had gone by since the mining disaster. And then it was the evening of Boxing Day - another Thursday. In fact it was around 10:30. Leo Comeau was on his way to work in the mines. His shift was the eleven o'clock one. As he passed the building occupied by

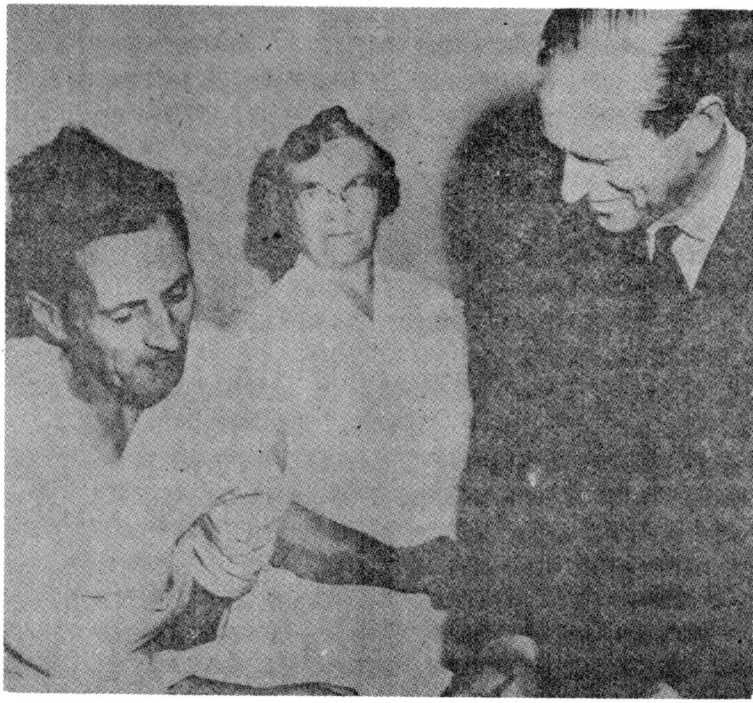

Prince Philip signs autograph for Levi Milley's daughter

The White Miner unveiled on September 11, 1894 by Governor General Lord Aberdeen and his wife, and the Right Honourable John Thompson, Prime Minister of Canada. This statue stands on upper Main Street.

The cemetery where many of the miners lie, including Henry Swift (see page 16). Among others buried here are "John Adams, Brakeman. A native of England, G.B., who was killed at Springhill by a train on the Springhill and Parrsboro Railway July 27, 1888, aged 24 years." and "James Elliott 1837-1926 Indian (Sepoy) Mutiny veteran."

Ronnie Skidmore, a boy of Canaan, near Springhill. Unlike some of the boys who were his age in the 1800's, he will never work in the Springhill Mines.

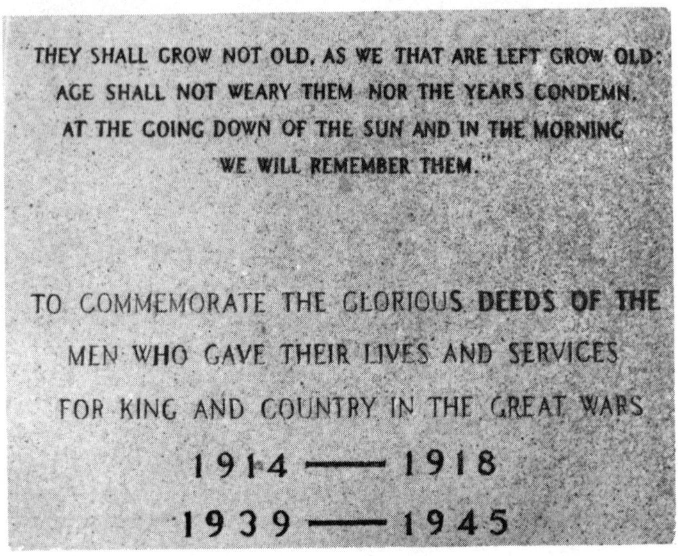

The War Monument near the White Miner

The entrance for the Number Two Mine as it is today. It is hard to believe so many perished below this now peaceful hollow in the earth.

Springhill — after the Great Fire of 1975. This was the third major fire to hit this town — the others also hit Main Street in 1957 and 1895.

The Town of Springhill (as seen from the old mine site) in 1976.

The present-day Miners' Museum takes people on a tour of part of the burned-out Syndicate Mine (burned in the late 1800's).

The author, Roger David Brown, on the grounds of the Springhill Miners Museum.

Stedman's 5¢ to $1.00 store and H.S. Terris Insurance Company,Limited, he noticed a fire burning in Stedman's basement. He didn't have time to ring in the alarm so he told Susse Porter. She immediately called in the alarm.

The Springhill Fire Department at first thought it could control the flames. Fire Chief Crawford told of the difficulties experienced at the Stedman's store. They couldn't get into the building from the front because the smoke was too thick.

"Even with masks we could not go in, because the masks are no good when the smoke contains less than 16 per cent oxygen."

Entering the building through the rear freight entrance they found flames everywhere in the heavily partitioned basement and going up above through the wall partitions. The Fire Chief was hit by a false ceiling which came down but he was dragged back to safety.

To add to the problem the winds came up and blew up to forty miles per hour that night. Help was requested from the nearby towns of Oxford (16 kilometres) and Amherst (27 k.). Shortly after this call for help, a terrific explosion ripped through the building - flames shot out the front into the buildings on the north side of the street. All control was lost. What caused this blast? Some felt it was caused by children throwing stones through the back windows of Stedmans which allowed the air to rush in and mix with the gases the burning store was producing.

Within a short time flames were racing through shops on both sides of main street. As the night progressed fire departments from Parrsboro (41 k.), Dieppe (in N.B. - 95 k. away) and Pictou (140 k.) arrived.

A two-hour electricity failure greatly hampered the firemen. This caused the booster pump in the water main to cease working which in turn caused a great drop in water pressure in the firemen's hoses. So much water was used that the pressure never built up to its proper level again that night.

Prince Philip and
Theodore Michniak

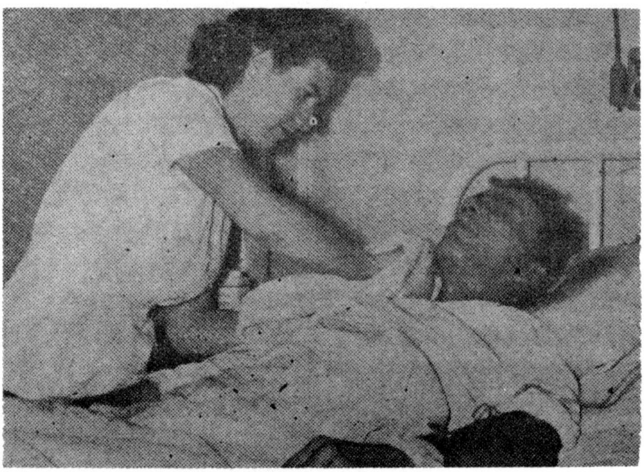

Garnet Clarke who spent
his 29th birthday underground

Adding to their troubles were three other alarms turned in - two of which were false. The first alarm was for Pettigrew's store on McGee Street one half mile away. This fire was apparently started by flying embers from the main conflagration. The store burned to the ground.

By five that morning the fire had been contained and a few hours later a heavy rain finished it off. But in that inferno 15 businesses had been lost and four houses on Fir Street were levelled. In addition 10 apartments above the businesses had been destroyed.

The damage was estimated at one million dollars; there were 35 people homeless and 30 more people to add to the jobless number.

This statement by Mayor Gilroy was recorded in the *Halifax Cronicle-Herald:* "We'll never say die. We'll carry on if only half Main Street is left."

Maurice Ruddick
"Citizen of the Year"

CHAPTER 4

"Oh, God — it was good to be out"

THE BUMP IN MINE NO. 2 IN 1958

Here at the base of the Spring Hill at 500 feet above sea level; that is, 150 metres, was the entrance to the only working mine left in Springhill, the number two mine. Many of those killed in 1891 had died in this mine.

Number 2 in 1958 was the deepest coal mine in North America - 1324 metres (4340 feet) straight down. The distance from the pit head to the end of the mine was 4362 metres - 14300 feet. Some of Dosco's mines in Cape Breton Island jutted out into the Atlantic six and a half kilometres - four miles - but number 2 was the deepest. The reason for its great depth was simple: it was old, it had first opened in 1873 when the mines of Springhill were really just starting. It closed in 1958. "I sometimes wonder if No. 2 is flooded now," Maurice Ruddick said to me. "I think it would be."

But on Thursday, October 23, 1958, the mine was in good working shape. People, including myself, were sitting in their houses laughing with Lucille Ball, one of the world's favourite entertainers. Her television programme was "I Love Lucy." Then, without warning, over a 15 mile radius the ground shook.

In Halifax, 119 k. or 74.5 miles away as the crow flies from Springhill, the instruments at Dalhousie University's Seismograph Station registered an underground disturbance centering on Springhill. Seismographs at Seven Falls,

Quebec and Ottawa agreed with the origin and on the time - 8:05:54 p.m.

The Royal Commission which investigated this bump gives a picture, which is easy to visualize, of the power of this disturbance. Imagine, it says, one thousand tons of coal being dropped forty feet onto the ground.

What had actually happened underground? In miners' language it is called a bump.

As coal is removed from the mine, pressure builds up on the walls and ceiling and floor from the overlying mass of rock. The deeper the mine, the greater the pressure. In the Springhill area the strata consists of strong material which does not readjust the pressure very gradually. Stress builds up when this happens. And the readjustment occurs by violent lurches as the weaker parts of the tunnel are wrenched, be it the floor or roof... These lurches can be set off in a mine by any slight underground change, from a fall in the waste to a tremor on the highway. To add to the danger, gas is released by the breakdown of the coal and the closing of approaches and ventilation systems.

One thing No.2 was famous for was bumps. In fact, said the Royal Commission, it had 525 bumps from 1917 until it was closed. For example, on October 9, 1956, and again on October 17 there were bumps in No.2. A bump was a familiar phenomenem in Springhill.

"Didn't you feel, " I asked Maurice Ruddock, one of the heroes of the disaster, "that a bad bump might hit?"

"Yes," he answered. "I told a fellow not long before it happened that a bad one was coming."

"Then why go into the mines?" I said to Mr. Ruddick.

"One thing you must understand," he said, "there were good times in the mines - a lot of singing in the break periods, a comradeship." And I know that people like to live in their own areas and in Springhill. In 1958 there was little work outside the mines.

As a rescued miner said, when asked if he would return

to the pits: "What else could I do? Coal is in my blood just like you can see it in the lines of my face. Sure I'll go back."

One miner told me he felt the cause of the bumps was that the tunnels were too close together for the strata to support the huge weight.

It was the afternoon shift that was at work that fatal evening, the same shift that had been caught in the 1956 explosion. They had gone to work with the usual routine. First, to the Wash House. There, from the ceiling, held by pulleys, hung buckets with long ropes attached. From a bucket a miner took out his work clothes and put in his street clothes. Newsmen, says the Halifax newspaper, were surprised to see clothes hanging from these ropes. The miners didn't have lockers. The rest of the buildings had showers. "Funny," said Mr. Ruddick, "hardly ever got a cold - would sometimes go out with wet hair from the showers."

Then there was the Lamp Building. Here the miners turned in a brass tag about the size of a silver dollar. It had a number on it. For the tag he would be given a lamp. The tag was then placed on a rack. This was the mine's numbering technique; simple but effective. When miners finished in the pits they brought back the lamps and picked up their tags.

Next, this shift had gone to the entrance of the mine and boarded the rake. It was a twenty minute ride down to the 13000 foot level. Here Maurice Ruddick and many others got out. Before him stretched the working area, a long shaft that had been bored into the coal seam. Then the miners had started working back to the main shaft, mining the coal as they went. Special groups of men pulled down the roofs in the places where they were finished. No electrical tools or explosives were used in this mine. People were well aware as to what a spark could do. Instead the machinery ran by compressed air. And the shovelling was done by hand. It was hard work.

The rake continued on to the 13,400 level where more men got off and then to the 13,800 level - the final one. The retiring shift then began to get on the rake for the trip to the surface

Nearly half the men, 80, worked directly on the coal face. A further 27 moved the coal to the slope and the rest were scattered from the levels to the surface running machinery.

At six p.m. the afternoon shift took a twenty-minute lunch break. And then they went to work again. Shortly before seven a minor bump hit the mine. It is a pity that it hadn't been stronger. If it had, the mine would have been cleared.

At 8:06, the moment of the disastrous bump, there were 174 men below in the mine. Seventy-five of them were to die.

After the shaking in the town stopped, the rush to the mine began. One of the first on the scene was the resident manager, George Calder. After several attempts he got through to the mine depths by telephone. The manager was informed that the main slope was in order but there was no communication with the sections below 7,800 feet. As McManamon said when he finally answered the telephone, "I guess I was knocked out for awhile. It's awful."

About 9:30 the volunteer rescue crew, led by Calder, reached the 13,400 bottom and found the level here blocked by methane. These volunteers had no face masks. Above ground the draegermenwere forming and entering the mine as the volunteers came up. The rescue work was efficient.

The first men up were Charles Alderson and Overman James McManamon. MaManamon and several of the men were chatting when it happened. "We just sat there for a few minutes - dumbfounded." He added that gas conditions were bad but there was considerable air down there.

A miner said: "The situation's terrible."

A rescue worker described the rubble in the depths this way: "It looks like a pile of spaghetti down there."

What were they doing, these lucky ones who survived?

Archie Legere, the man who led the trapped miners in 1956 in singing "Faith, Hope and Charity", had been at the 13,800 foot level. His life had been saved by an accident: instead of moving up the coal wall he had gone to eat.

Another survivor was Bill Totten, a miner for 32 years: "All I know is I lost my light and I lost my uppers. I don't know what I'll do, maybe I'll be a commissionaire or something. I've always wanted to do something like that. It will be better for me and better for my family." He had been operating an engine at the 13,000 foot level, 90 metres from the coal wall.

"When it came I didn't know what happened. Something from the machine hit me in the head. I found my lamp, got it working, and went back to what was left of the engine, and turned off the air valves because the lines were screeching like crazy.

"I moved along a piece and met four others, Buddy Pyke, Jackie Newcombe, Ronnie Cameron and Raymond Murphy. We groped along as best we could, keeping our heads down because we were afraid of the gas. You could hear her snapping and cracking, rocks and dust were falling everywhere. The level is usually about 12 feet square, but we had to crawl through an area about two feet in diameter to reach better air and space.

"We tried the phones on the 13,400 level but couldn't get anything. It was terrible. We couldn't even see our foot tracks where we had stepped. The dust filled them as quickly as we made another. We walked up the back slope and into 13,000. The air was bad. We met a man crawling out on his hands and knees using the rail as a guide and with no light. He was cut and bruised (Cameron Fraser). He wanted water. We gave it to him and rested as long as we could and went back to the bottom to try other levels.

We got to the 11,800 level and finally the rake came down for us."

The *Halifax Chronicle Herald* on Friday morning

carried the news that 72 men had been brought live from the mine and nine more were known to be dead. Eventually the rescued reached 81.

On Friday the pit-head was almost deserted. The description of the upheaval by the miners led most people to lose hope and the families and friends slowly walked away from the pit and went home. A few stayed with forlorn hope.

"I don't think there is much hope for them," said one 14 year old boy when speaking of his father. There wasn't; his father was dead.

Another boy said: "I wouldn't be surprised if they were all done for. This is sure an awful way to get a holiday from school. I'm sure glad my father doesn't work down there."

And Gary McDonald said of his father, "I have been through this before. I don't know what to expect. I'll just sit and hope and pray he gets out, I guess." He did.

Through the Governor-General there came a message from Buckingham Palace - just as one had come in 1891. The Queen was distressed and wished to be kept informed.

Saturday General Manager Harold Gordon of Dominion Steel and Coal Corporation said that there was no hope for survivors in the 13,400 and 13,800 levels and little for those at the 13,000 level.

Then the days rolled by and no survivors came up.

Unknown to those above ground, at the 13,000 level two groups had formed. Their stories gave to Springhill its twin miracles.

The rescue parties had laborious and frustrating slow going. Taking the path of least resistance, they dug with sawed off shovels snake-like paths in the debris. These tunnels went up and down, left and right through the devastated passages. The men dug on their hands and knees and in the most advanced parts of new tunnels they were working while lying down.

"One of the biggest problems," said a draegerman, "is

the debris - there is no place to put it, but you have to lose it somewhere."

And then there was the gas. To combat this, auxiliary fans were installed to draw the bad air into ten inch pipes.

Slowly twelve men formed a group from the ruins of the 13,000 foot level. Here is their story as they told it themselves.

On that Thursday at 8:06 p.m. Harold Brine had been coming from the pans when all Hell broke loose. When he become conscious again he could hear voices down the wall. Slowly he began to move. He found a man half buried in the rubble. "Harold Brine dug me out after the fall," said Hugh Guthro later. "I wasn't hurt much." He was lucky.

Thirty-nine year old Joe McDonald was hurt though. "I fell heavily to the floor and rocks landed on my left leg and it hurt something terrible. It wasn't long before some of the boys - I'm not sure who - came over to us. They tried to move us but it hurt too much. They propped me against the wall and I had to sit like that for six days."

The other man McDonald had referred to was Theodore Michniak. The two had been working close together. Michniak had various wounds including a dislocated shoulder.

"I don't know why I wasn't killed," Wilfred Hunter told reporters. "The boys were dying all around me. The groans soon died out. I don't know how long it lasted. I guess I was stunned...We called to each other counting noses finding out our names...We crawled around on our hands and knees."

"Most of those who weren't killed instantly by the bump died shortly after from the gas," said Caleb Rushton later. "The gas wasn't too bad at first and later we got some air to clear it out. But those who had been injured didn't have the strength to fight it off."

Gorley Kempt, after the cave in, had taken a lamp from a dead miner and with this he had found his way down the mine tunnel to where the others were forming the group.

Harold Brine ate the last of his food that Thursday night. After that he joined the others in looking for extra food cans, and in looking for a way out. There was no exit. In one of the searches Wilfred Hunter said later, "I reached up and touched a body pinned by the rocks. I could tell by feeling that it was my twin brother, Frank. There was nothing I could do." He said that he told Frank he was sorry, so sorry that it had happened.

"I don't know how much water we had," added the wounded McDonald, "but it was not near enough. It was the thirst that was the worst. I didn't sleep too much at first and the thirst bothered us a lot."

"We had a two gallon can of water. Later another can was found," said Wilfred Hunter.

Kempt told of his six days, "I dug a cubbyhole in the rubble. I thought there might be a way out but there wasn't. So I just lay there day after day. And I prayed - and I'm not a praying man."

"It was while we were nibbling on our few sandwiches that I started them singing hymns," said Caleb Rushton. "I'm a choir singer. The boys really liked it when Bowie Maddison and I would do a duet on 'The Stranger of Galilee'. That was their favourite and we sung it many times each day. Prayer was what kept us alive". With a luminous watch he kept track of the hours which became days.

Hunter mentions, "Sometime Friday we heard a scratching noise underneath us. It only lasted an hour and we never heard it again. It must have been another miner on the level below us. We really started getting thirsty on Sunday. My tongue started to stick to the roof of my mouth. We never gave up hope. But by Tuesday night we were terribly thirsty. We slept most of Tuesday night and Wednesday morning, and started tapping again Wednesday - I think it was morning."

Meanwhile their group had been increased from

eleven to twelve. Joe Holloway was the last to join the group. He had been at the 13,000 foot wall when the bump hit. Badly dazed and bruised he suffered from gas as well. For three days he had hunted for a way out and then he had joined The Group.

On that Wednesday Gorley Kempt was lying with his mouth against an air pipe - the pipe where they had been tapping their signal. He heard the clink of metal hitting metal. "I couldn't believe it when I heard that cable rattle. I wouldn't let myself believe it. I just lay there listening. Then I heard a pick or something click on the pipe and I hollered. God, how I hollered...and when someone hollered back - well I just can't describe the feeling."

Earl Wood, a veteran of 31 years of mining described the reaction of the rescue workers: "I was working away in a shaft leading from the main slope, digging in a solid wall of coal when all at once the coal became very loose and, with a roar, a cloud of dust and hot air blew into my face. This was about 11 o'clock and we sat there half an hour or so," said the 48 year old bare-faced miner. "We were hopeful because we felt sure we had reached the opening we were hunting for. There was some gas and it was terribly dusty.

"I went back again with two of the officials. We had high hopes that there might be a hole big enough for a man to crawl through. All we saw was a stump of a six inch steel compressed air pipe. It was broken off and we could not see or hear anything. After we had been sitting and watching for some time we thought we heard someone speak. It was just a piece of a word and very faint but it hit us like an electric shock. We looked back toward the loader, which was on the way back to the slope, and found about 40 feet away that the pipe was broken off."

The *Chronicle-Herald* account continues: "We figured that the voice we heard must have been some one out in the tunnel and we sat down to wait. About 2 o'clock the superintendent and the engineer decided to check the air coming from the pipe I had uncovered. The reflection of the the bottle must have sent light into the pipe. Then had come Kempt's words, "We are alive in here - how about some water?"

Wood continued, "Two of the fellows with me, Arnold Reese and George Hodges, had a brother-in-law, Harold Brine, in there, so we let them do most of the talking. They still had time for jokes."

When George Scott, a man with the Scottish accent, yelled to them, Kempt joked: "Take the marbles out of your mouth - talk English."

McDonald tells of the last days of the entombment, "Time seemed to travel so slowly. We were tapping on the wall most of the time because we were sure we could hear digging nearby. Finally, just before noon, Wednesday, someone said, 'My God, we're saved.' Gorley Kempt...hollered. I heard a voice on the other side and I knew we were saved. I even broke down and cried a few times. Thank God I am alive. Twice (trapped in '56) is enough. I will never go down again."

"After we made contact," said Wilfred Hunter, "we got soup and water (through the tube). It wasn't much but it was the best meal I ever had. Time seemed to go quick and soon the boys had beoken through.

"They cried a bit and we did too... We were all laughing and crying all at once as they brought us to the surface. They put a blanket over my face but I showed them I wasn't finished. I took the blanket off and waved to the crowd. I remember that and I remember how they cheered. Oh God, it was good to be out."

And the twelve came out:

The first - GORLEY KEMPT ("I was cleaning up after all the company we'd been having," said Mrs. Kempt, "when the bulletin that 12 had been found was flashed over the television. We just sat there waiting for the names. We didn't think we would get them for awhile. I almost couldn't believe it when they said Gorley Kempt had spoken to the rescuers. It was lucky for me he happened to be the one.")...HAROLD BRINE...JOSEPH HOLLOWAY...WIL FRED HUNTER (Mrs. Hunter hadn't picked up her husband's pay check as some wives had done; she hoped. She told her husband; "I thought I'd leave it for you to pick up yourself.") ...LARRY LEADBETTER ... LEVI MILLEY ... THEODORE MICHNIAK ("I just can't find words to tell you how it feels to receive the news. I've been waiting for word ever since the accident. He's a real good man, " said Mrs. Michniak, "Thanks to God for a miracle.") ...CALEB RUSHTON ... BOWMAN MADDISON ("I just can't say anything yet ... I've just learned the news. I just felt like getting up and running to the mine. But then I had to feed my four month old baby," Mrs. Maddison said.) ... ELDRED LOWTHER (Bobby Lowther, 11, said, "Oh, boy, now we'll play baseball again together. I could hardly believe it was true when I came from school and my mother said 'Daddy is alive'. I was so happy I cried. My daddy and I used to go out in the backyard every day after supper and play ball. He made me throw the ball over the plate; now I'm a pretty good pitcher.") ... JOSEPH MCDONALD ... and HUGH GUTHRO ("I don't care if we starve to death," said Mrs. Guthro. "Hughie will never set foot in another mine. There must be better things to do. Everyday he went to work I worried. He didn't like working in No. 2. I don't think he was afraid of No. 2 but he didn't like it.")

The first miracle was over. Among the searchers was Percy Hunter. He had been trapped at the 13,400 level and been found by Calder and his men. Since that time Hunter had been searching for his twin brothers, "the boys". Now he had found Wilfred, alive. And Wilfred said Frank was dead.

The jet plane bound for London from Ottawa touched down at the Moncton, New Brunswick airport. From this aircraft emerged His Royal Highness, the Prince Philip. He and his wife, Queen Elizabeth II had been to Springhill

Junction before on a triumphant tour. Now there was tragedy. Eldred Lowther, who had spent six days entombed, expressed the feelings of the people when he said, "Thank you for coming to see me".

Meeting the Prince at the airport were several local leaders: Major General Plow, Lieutenant Governor of Nova Scotia; Premier Stanfield of Nova Scotia; Federal Revenue Minister Nowlan and Springhill's own Stephen Pyke who was Nova Scotia's minister of Labour and Public Works.

The two car convoy sped toward Springhill. Once in town the Prince went straight to All Saints Hospital. "Boys, meet His Royal Highness, Prince Philip," said Mayor Gilroy.

"It's good to see you," said Levi Milley - a man noted for his humour. Milley told his prince that he had pulled another fellow out in the mine.

"Was that how Mr. Stevenson's leg was broken," answered Philip, jokingly referring to the man in the neighbouring bed.

Autographs are a royal family taboo, so it was almost unprecedented when he signed a piece of paper for Judy Milley - the daughter of this man who had spent six days underground.

"How long have you been a miner, Mr. Michniak?"

"Since 1918," came the reply.

At the armouries - which had become an emergency hospital - came a second autograph. George Hayden, one of the first to come out of the mines, was lying with a cast on his left leg. When the prince left, on the cast was the signature "Philip". Hayden's wife was almost speechless: "I can't remember anything he said."

At the mine Dosco General Manager Gordon outlined the general situation, showing plans of the rescue operations. Don Ferguson, his cheek still marked with a gash, broke through the crowd, grasping Philip's shoulder. "I'm glad to meet you sir. I'm glad to see you here. Just seeing you has done me a world of good."

Said the Duke, "I'm glad to meet you, too."

Inside the pithead the Prince looked down the slope where the rescue work was going on. Dosco had said earlier that they would not allow him to enter the mine, because there was a danger of new collapses. When he emerged from the mine the pleased barefaced miners cheered.

The final visit was to the home of Mrs. Harold Raper, a woman whose husband had been killed. He had been a member of the Order of the British Empire. "He only stayed about ten minutes. I told him there should be a padlock on that mine down there. I told him all we want here is an industry that will keep our men out of the mines because this is a good town with a lot of good people in it." She didn't give the Prince's answer because "it wouldn't be right". She said of the man for whom she had made a cup of tea: "He wasn't like a royal prince but a prince of a man. He made us feel so easy: he was like a friend we hadn't seen in a long time - he was lively."

After two and one half hours in Springhill the automobile with the royal standard flying headed back toward Moncton. It was All Hallowed Evening - Hallowe'en - the night before All Saints Day. From the car the prince could see along the roadside the children dressed as the witches, ghosts and goblins and all the other creatures we have inherited from the Middle Ages.

At the New Brunswick border Lieutenant Governor O'Brian met him and in another hour the airplane flew off toward London.

Underground, unknown to the searchers, there was still another group to be found.

On that Thursday, when the bump came, 35 year old Herbert Pepperdine was thrown about six or eight feet downhill at the 13,000 foot level. "The first thing I saw when I came to was Garnet Clarke - laying on top of me. We started down. Then Garnet, Frank and I helped Doug Jewkes out from among the stone and pack."

Jewke had been thrown into the air, then smashed

onto the floor and half buried: "I'm so sore," he said later, "My hips are all black and blue. My battery pack was driven into my hip."

The Frank, Pepperdine had referred to, was Frank Hunter - the missing twin. He wasn't dead as his brothers thought. This was the third calamity to befall him since he had started working in the mines. In 1952 he had been half buried under a fall of coal and had had his back broken. In 1956 during the explosion he had been trapped four days.

"At first, right after the bump," said Maurice Ruddick, "we got gas. We went woozy - and I thought we'd had it. Then it cleared up. When we tried to go up the timber road we had to go on our stomachs, because the road was right up to the roof ... then we had to come back, crawling; you talk about a job!"

Slowly a group of seven had formed. Among them was Percy Rector, a man who had been entombed for nearly two days in the 1956 blast. Rector's right elbow was jammed tight in a smashed wood pack. There was only one way to free him - to amputate the arm. The group chose not to do this. A rescue party might reach them shortly and the operation itself would be extremely difficult to perform down there. The only medication they had for him was a few aspirins.

The day after the bump Pepperdine, Smith and Clarke found Byron Martin. He was in a hole between stone packs. "I moved Barney, he seemed to be dead," told Pepperdine. "There was a nice breeze coming through a place that was partly cleared. He probably did it himself, so we made the hole bigger to let more air up." Since they had no lights and they would have to drag Martin, they thought it better to leave him where he was. So he remained thirty metres from the group, the eighth person in the tomb.

Ruddick, the singing miner, sang hymns. "I got along pretty good at first, but later the singing got harder. My voice cracked because I was so dry," he said.

And Garnet Clarke told, "It was very hot."

"What did you drink?" I asked Ruddick.

"We had some water - and later on we drank urine. If we had only known it, by a dead body down there was a container with a fair bit of water."

Pepperdine told of the danger of the waste falling in, "One time we gathered up as much timber as we could and built a pack to make it safer." Another time, he said, "I went scrounging for food. I had no light and I had to kind of feel around in the dark." He found a ten cent chocolate bar: it supplied seven men with two meals.

On Monday, after three and one half days trapped, they held a birthday party for Garnet Clarke. He was 29. The birthday cake was a sandwich cut into seven pieces, one for each man including Percy Rector. For a drink they each had a bottle capful of water. "Happy Birthday, dear Garnet, Happy Birthday to you," they sang.

Tuesday the water ran out. For moisture they chewed the moist bark from some props and some chewed a pinch of coal. As Currie Smith said it helped to take away the gnawing in the stomach. He also drank a bit of oil: "It didn't seem to hurt me."

"It was just terrible, terrible," Jewkes said later. "We were like trapped rats, worse than that ...we could hear the rats. We were afraid they'd eat us."

They pounded on the air pans and pipes. "We wore ourselves out doing it," said Ruddick. After several days they heard the rescuers' chipper pick. When they didn't hear it they were afraid the searchers had given up. Then would come the sound of the distant pick again and with it renewed hope.

On Tuesday Percy Rector stopped moaning. He was dead.

Above ground Mrs. Herbert Pepperdine was standing in the open with her friend. It was the time of the miracle when the twelve had been brought out. The women waited. "There's always hope in a mining town," one said.

On Friday, eight days after the bump, Smith and

Clarke paid their third visit to Martin. He had moved a bit. Smith said, "We were too weak to move him, and he was too weak to go out alone. He asked us for water but we had none."

"I wasn't really scared. But I thought we were going to die," said Jewkes. "I couldn't have lasted another twelve hours. My tongue was thick - so thick."

"We were all lying down," said Ruddick, "when we heard John Calder hollering, 'How many here?' "

"He sounded just like an angel," said Jewkes.

The eight and one half days of hell were over. And the first one out was Byron Martin. One of the rescue workers said, "I guess Byron Martin had it the worst of all. He spent the entire time all alone." Said another worker, "The first thing we heard were moans. We dug a little more-there was scratching." Martin himself mumbled, "God must have saved this little hole for me." Later he said, "Thank God I'm alive ... I didn't think anything could be so bad."

"When we found them, there was Maurice Ruddick, sitting on a stonetack, singing at the top of his lungs. He's called 'The Singing Miner' and I guess he sang," said James Rossong. "The other boys said he was the spark plug that kept up their hopes." Said Mr. Ruddick, himself, to me, "When they came I guess I said, 'Give me a drink of water and I'll sing you a song.' "

The joy of the relatives was unbounded. Said Mrs. Martin, "I was never up at the pit. I knew I'd just break down and cry. I was up half the first night just looking out the window toward the mine. I never gave up hope." When the twelve were brought up she "felt shaky. I just broke down and cried. I felt just so happy I couldn't feel any better."

Herbert Pepperdine's wife fainted.

Maurice Ruddick's wife, the mother of his 12 children, said, "I was still in bed at the time and I asked my sister to see who was at the door. It's funny, but I'd been thinking all along how scared I'd be if I saw the minister coming, but I wasn't a bit scared when I heard the knock: I didn't think of anything. I just called to my sister to answer the door. When the minister told me Maurice was saved, I grabbed him and hugged him, then I ran around telling everybody in the house. I still don't quite know what I'm doing."

Currie Smith's wife, Mabel, said that when she heard the news, "I just leaned on my refrigerator and cried, until I couldn't stop."

It was over. The death toll stood at seventy-five.

The 1958 disaster fund would eventually reach two million dollars.

INTERLUDE

The number two mine never reopened. Hard times came to Springhill. People, hard pressed, went to the great slag heap to gather fuel for themselves. Some lit fires, which burned among the piles of coal and after awhile proved impossible to put out. Near the present museum these acres of coal can be seen and smelled burning. The Government estimated it would cost a million dollars to put out the fire.

The population of the town dropped to the five thousand level, little more than what it had been in 1891. Near the present museum the Syndicate mine opened and employed a relatively small number of men. In three accidents three men lost their lives. In 1970 the mine closed.

And that was the end of mining in Springhill.

Fortunately for the town several small factories moved in since 1956. This did stop the town's downward slide and it is now recovering. The new schools and new stores bear witness to this. As someone once said, "There'll always be a Springhill."

In the late 1960's and early 70's Springhill produced two excellent musicians - Johnny Mooring (a North American fiddling champion, killed in Montreal in 1974) and the world famous singer Ann Murray. A friend of mine, Conrad Bayers, was in Central Asiatic Russia in the early 70's in the large city of Irkutsk. As a foreigner he was invited to join a dozen Russians at a table, who were celebrating a comrade's birthday. When one of these students found out he was a Canadian the question was asked, "Do you know Ann Murray?"

EPILOGUE

And then on July 20, 1975, came:

Another Spin of the Wheel

Main Street wound its way up Monument Hill, as it had for a century: the lower half consisted of wooden buildings, crowded together, and here and there a non-wood structure, like the fine old town hall with its high brick clock tower (a similar structure stands in Parrsboro). After these buildings came spaces left from the 1957 fire, and further on up the hill came new shops and then the aged churches...

A strong wind swept the Hill early on this Sunday morning. At 2:45 a.m. two yound men noticed a fire burning in a lower Main Street restaurant. These two, Ronald Moore and Glenn Noseworthy, gave the alarm. Fire departments poured in from all over Cumberland and the adjoining counties of Westmorland and Colchester - fourteen fire departments in all would battle the blaze.

The list of places destroyed was appalling - the town's only cinema, the Capital; its Town Hall, Police Station...shops and residences, the new *Springhill Record* office...Mrs. King's home on Church Street was dynamited to stop the flames. A total of eighteen buildings on both sides of Main Street were burned. The conflagration finally was ended by the lots left barren from the 1957 fire.

By 7 a.m. the 130 firemen had the fire under control, but it smouldered for another week.

The *Halifax Chronicle-Herald* on July 22 recorded Mayor Mont as estimating the cost of rebuilding at as much as ten million dollars.

And then Springhill started to rebuild. Good luck to her.

APPENDIX

"Some people say that the 1891, 1956 and 1958 disasters were the great disasters," said the Springhiller, looking at a list of the dead in the mines. "But for me, this was the great disaster," he said, pointing to a solitary name, the only one listed for that accident, "that is my father."

Below is a list of the 424 men and boys who lost their lives in the Springhill Collieries since January 1881.

	DATE	PLACE	NAME
1881	January 20		BLUE, John
	October 6		BIGNEY, Robert
1883	December 20		FLETCHER, W.
1884	August 20		CROCKER, Fred
1885	August 28		MacDONALD, Donald
			SOULLY, John
			HOSLEM,...
	December		WILSON,...
1886	May 3	No.1	MacLEOD, Murdock
	November 3		TURNER, George
1887	February 7		MacDONALD, William
	April 2		MAY, George
	June 25		MACKIE, James
	December 8		GUTHRO, Thomas
			CRAWFORD, Clinton
1891	February 21	No. 1 Seam EXPLOSION	ARMISHAW, Herbert
			ARMISHAW, Jesse
			BAINBRIDGE, Ernest
			BENTCLIFFE, John
			BOND, George
			BOYD, John
			BROWN, William
			BUDD, Alonzo
			BUNTON, Alexander

BUNTON, Andrew
BURCHILL, William
CAMPBELL, Alex
CAMPBELL, Donald
CAMPBELL, John D.
CARMICHAEL, Andrew
CARMICHAEL, John
CARMICHAEL, William
CARRIGAN, William
CARTER, Clarence
CARTER, Reed
CARTER, Willard
CASEY, Jude
CHANDLER, Ernest
CLARKE, Robert
COLLINS, Mathew
CONORTON, John
CONWAY, James
CRAWFORD, John
DAVIS, Thomas
DAWSON, Richard
DAWSON, Samuel
DILLON, Frederick
DUNN, John
DUPEE, Joseph
ERNEST, Roger
FIFE, Hiram
FINDLAYSON, Daniel
FLETCHER, Thomas
FRANCIS, John
FURBOW, Samuel
GALLAGHER, Peter
GILLIS, John
GUTHRO, Lazarus
HALLET, Thomas
HANNAGER, Peter
HAYDEN,John
HUNTER, John
HYDE, William
JOHNSON, James
KENT, William
LEGERE, Samuel
LETCHER, Frank
LIVINGSTONE, Henry
LOCKHART, Daniel
MacDONALD, John J.
MacDONALD, Rory B.
MacEACHERN, John D.
McFADDEN, Robert
MacGILLIVRAY, Will
McKAY, Donald, 2nd
McKAY, H. J.
McKEE, William
MacKINNON, Allan
MacKINNON, Angus
MacKINNON, John
MacKINNON, Laughlin
MacLEOD, Henry
MacLEOD, Neil
MacLEOD, Norman
MacLEOD, Rory
MacNEIL, John P.
MacNEIL, Rod C.
McNUTT, Charles
McPHEE, Neil
McVEY, David
McVEY, James
MAIDEN, William
MARTIN, George
MILLER, James Sr.
MITCHELL, John
MORRIS, James
MORRISON, Thomas
MOTT, Ernest
MUCKLE, Samuel
MURPHY, Jeremiah
MURPHY, Richard
NAIRN, James
NAIRN, John
NAIRN, Malcolm
NASH, Charles
NICHOLSON, Malcolm
NOILES, Roger
OVERS, James
PIQUINOT, James
PITT, Joseph

REGAN, Bruce
REID, Peter
RIPLEY, Clifford
ROBBINS, James
ROBERTSON, Hugh
ROGERS, Thomas
ROSS, Philip
RUSHTON, Stephen
SHARPLES, James
SHERLOCK, Robert A.
SHIPLEY, Archibald
SIMMONDS, Howard
SMITH, Edward
SWIFT, Henry
TATTERSAL, Joseph
TAYLOR, Douglas
TURNER, William H.
VANCE, Alex
WATT, David
WHITE, Philip B.
WILLIAMS, John
WILSON, Thomas
WOOD, George
WRY, Edgar
WRY, Henry

Year	Date	Location	Name
1892	October 3		HOPKINS, George
	October 18		COSTIGAN, Richard
1893	November 18		TEBE, James
1894	February 26		FOSTER, Alex
1895	January 16		GUALTIERE, Natali
	February 26		HORONE, William
	August 31		MATHESON, H.T.
	October 8	No. 3 - 2600 l.	MILBURN, William
1896	February 24		GALLANT, Lawrence
1897	May 29		BURROWS, John
	September 28	No. 3 2600 lev.	CUNNINGHAM, William
	December 8	No. 1 Pipe Slope	FERGUSON, James
1898	February 17		LeFAVOUR, Emil
	May 12	No. 3 mine	BORAN, John
1900	April 14	No. 3 2600 lev.	CARGIG, William
	May 22		LOWTHER, William
1901	February 28	No. 2 Pipe Slope	FURLONG, Frank
1902	April 14	No. 3 mine	SWEET, E.
	June 12	No. 3 mine	BURNS, P.
	November 24	No. 3 mine	WILSON, James G.

1903	February 4		FISHER, Henry
	April 1	No. 3 2600 level	PURDY, Charles L.
	October 28	No. 3 mine	HYATT, Percy
			McMILLAN, John
1904	January 30	No. 2 West 3300 l.	ELBRIDGE, Clifford
	August 5	Aberdeen mine	FOSTER, Arch
	December 31	No. 2 West 3300 l.	BRADLEY, Ephriam
1905	April 3	No. 2 2400 level	HANNAH, Joseph
	January 16	No. 2 2400 level	SCOTT, John
	August 29	No. 3 bump	MacKINNON, A.
	September 11	No. 2 2400 level	CHERRIE, William
	December 2	No. 3 2600 level	HENDERSON, William
1906	May 28	No. 2 Upper Screens, R'road	ELKIN, Fred
	June 19	No. 3 - Trip 2600 level	DAVIS, Ben
1908	January 30	Fell in Chute	DAVIDSON, Adam
	May 8		RIPLEY, Edward
	August 13	No. 3	JORDON, Perdy
	December 9	No. 3 2600 level	FOSTER, Arch
	December 12	No. 3 Bump	BIGNAY, Alex
	December 21	No. 3 Main Slope	WEATHERBEE, Melvin
1909	January 22	No. 3 Chutes	DELANEY, Douglas
1910	August 26	No. 1	KRANTZ, Joseph
1911	August 31	Surface Stone Trussel	GILROY, Joseph
	December 14	Surface Coal Car	PAUL, Percy
	December 31	No. 3	BULMER, Walter
1912	March 22	No. 3 Bump 2600 level	MacKENZIE, Hector
			GRAHAM, Jabaz
	December 18	No. 2 Main Slope	JONES, Moses
1913	August 12	Main Slope 4700 l.	GILLESPIE, William
	November 1	No. 3	CLARK, Ernest
1914	January 11	Main Sl. 2800 l.	HACQUOIL, Alex
	December 23	West Slope	CASEY, Percy A.
			CAULFIELD, Peter
1916	November 13	Main Slope 3200 l.	GILLESPIE, David
	November 23	No. 2 Screen, R'road	KNOBI, Fred

Year	Date	Location	Name
1917	September 14	No. 2 Bump 5400 l.	PIGGOTT, Richard
	October 17	No. 2 Bump	STEVENSON, William
1918	January 16	4700 Bull Wheel	THIBODEAU, Godfrey
	March 25	No. 2 Bump 4700 l.	GABRIEL, Alex
	August 30	No. 2 4700 level	JERROLD, Clement
	September 12	No. 2 Fanway	RONNIE, Ottie
1919	February 10	No. 2	TERRIS, ASA
	February 19	East 5400 bump	TURNER, Joseph
	April 19	No. 3 Boiler Shed Surface	TAYLOR, John B.
	June 12	No. 2 Main Slope - 5400	BLUE, Duncan
1921	September 3	5400 East	CASY, Archie
	October 12	No. 2 mine	DUNPHY, Carl
1922	March 8	No. 2	JONES, Hezekiah
	August 4	No. 3 Main slope	MERLIN, John
1923	January 8	No. 7 1000 level	HARROUN, John
	March 9	No. 2	MELANSON, Melos
	July 18	5400 West Bump	DAVIDDONKOW, Pichon
	September 25	5400 West Bump	BOURGISK, Charles
	December 20	5400 West Bump	REID, William
1924	March 20	West 5400 level	MacPHERSON, Hugh
	August 4	East 5400 Bump	BARRETT, Victor
			PETTIGREW, Charles
	December 6	East 5400 Bump	SWEENEY, John
	December 12	West 5400 Bump	KEIVER, James
1925	February 17	No. 2 Killed by Trip	BOSS, Asa
			MITCHELL, William
1926	February 3	East 5400 in Pillar	BROWN, George
	July 2	5400 West Longwall	MILLER, Penman
1927	May 17	No. 7 - S. 1500	DICKSON, William
	December 27	No. 6 - S. 1500	WILLIGHER, Mark
1928	July 3	No. 2 - 5700 East Bump	PENNY, William
	September 17	No. 2 6300 level	MEAGHER, Milton
	November 22	No. 2 6500 level	MOLNAR, Steve

Year	Date	Mine	Location	Name
1929	January 19	No. 2	7100 level	NELSON, William
	August 13	No. 7	2000 longwall	SCHWINGHAMMER, Fred
	November 26	No. 6	1500 level	BURTON, John E.
1930	March 11	No. 6	2000 Longwall	MORSE, Harry
				FAGON, Edward
1931	March 9	No. 2	7100 East	CZALONIC, John
	December 23	No. 2	Bankhead	VANCE, Alex
1932	February 29	No. 1	4400 West	DYKENS, Elmmar
		No. 7	2000 Longwall	YORKE, Eugene
1935	January 17	No. 2	Back Slope	CARTER, Ralston
	February 26	No. 4	Back Slope	BRYAN, Joseph
	October 22	No. 2	7100 level	FOWLER, Orrin
1937	March 20	No. 2	8300 Trap Door	LUTES, Norton
	June 15	No. 4	2000 Longwall	BUDD, Walter
	July 22	No. 4	2000 Longwall	MAIN, Douglas
	November 8	No. 4	Sinking Slope	WEATHERBEE, Albert
1938	February 11	No. 2	9000 West	TABOR, Fred
1939	March 24	No. 1	Fan Slope 400	TYOPENHART, Joseph
	May 18	No. 2	8300 Longwall	ROSS, Edward
1940	March 29	No. 4	4400 West Btm.	VIENNEAU, John
1941	January 30	No. 1	4100 West	HUNTER, Joseph
	July 15	No. 1	5700 East	GILLIS, Frank
	October 6	No. 4	5100 level	O'BRIEN, Thomas
	December 2	No. 2	10200 level	NODWELL, Harry
1942	May 29	No. 1	4700 Tunnel	WEATHERBEE, Lorne
	July 24	No. 2	9800 level	CLARKE, Walter
	October 14	No. 4	5700 level	STEVENS, Freeman
1943	January 11	No. 2	9800 level	WEIRICH, Herman
	January 23	No. 2	Mine Bank & Screens	FOWLER, Richard
	May 19	No. 1	3800	LEWIS, John
	July 20	No. 4	5400 level	MICHNEAK, Emmerson
	November 20	No. 2	9000 level	MEGEHEY, Vernour

1944	January 29	No. 4 5400 level	JACOBS, Louis
	March 4	No. 4 5400 level	McKAY, Daniel
	March 13	No. 4 6000 level	CHAPMAN, Neil
	July 12	No. 2 9400 level	RUSHTON, William
	August 1	No. 4 5700 level	CALDWELL, Charles
	August 30	No. 1 5100 level	NEWMAN, George
1945	August 27	No. 2 7100 level	PORTER, Thomas
1946	August 26	No. 1 5100 level	PASHKOSKI, Alex
	September 7	No. 4 6000 level	COCHRANE, William
1948	April 8	No. 4 6350 level	LIGHT, Fred
	June 29	No. 2 2800 level	McGRAW. Ernest
	June 29	No. 2 Pumphouse	LEGGATT, Herbert
1949	March 2	No. 2 11000 level	GABRIEL, Thomas
1950	June 8	No. 4 7100 level	CROSSON, Harry
	November 7	No. 2 11,400 level	MOSS, Laurie
1951	September 14	No. 4 7400 level	ALLEN, Arthur
1952	January 14	No. 2 11000 level	GOODE, Allan
	March 11	No. 4 7400 level	NELSON, Vernon
	April 16	No. 4 5100	SKILTON, Hollis
	May 25	No. 4 8100	CANNING, William
	July 2	No. 4	HACHEY, Archie
	November15	No. 2 - W. Pipe Slope	DAVIS, Andrew
			BRENNAN, James
1953	January 13	No. 2 10600 level	ROSS, Thomas
	February 13	No. 1 5100 level	BELL, Thomas
	April 7	No. 1 5700 level	NICHOLSON, Thomas
	August 25	No. 2 11800 Bump	PAUL, Charles
			DAVIS, Douglas
	September 1	No. 2 6900 level	CASEY, James
	October 23	Surface	COTTENDEN, William
	October 29	No. 1 5100 level	WELCH, Clarence
1954	July 29	No. 4 Back Slope	GILROY, Russell
			KNOWLTON, Everett
			WHITE, Leo J.
	November 2	No. 2 Main Slope	BEATON, John

Year	Date	Location	Name
	December 8	No. 2 11800 level	HUNTER, Wesley
			DAKIN, Layton
			FEAR, John
			LORIMER, Harold
			COPELAND, James
1955	September 22	No. 2 Fan Slope	REESE, Lloyd
	November 22	No. 2 Sinking	McSEPHNEY, William
	August 17	No. 2 Screens Bank	BOUTILIER, Gerald
1956	January 27	No. 4 6100 level	HAHNEN, John
	May 2	No. 2 12600 level	WHITTLE, William
	May 31	No. 4 6100 Slope	HALE, George
	November 1	The Explosion	CRUMMY, Joseph
		No. 4 Bankhead	JONES, William
			MacDONALD, Leslie
			MacLELLAN, Ben
			NELSON, Lester
			PYKE, Pleaman
			VANCE, David
		No. 4 Draegerman	FERGUSON, William
			SPENCE, Alex
		No. 4 Explosion	ANDERSON, Ephraim
			ALLEN, Frank
			BEATON, Floyd
			BEATON, Kenneth
			BETTS, David R.
			BOUTILIER, Ernest
			BROWN, Thomas
			CAMPBELL, Alex
			CLARKE, Kenneth
			CLARKE, Ralph
			DAKIN, Gilbert
			DAWSON, Gerald
			ELLIS, Richard
			FISHER, Lester
			GLENNIE, Avard
			HENWOOD, Victor
			HUNTER, Angus
			LEWIS, Harold

		MacLEOD, Henry
		MILLARD, Victor
		MILTON, Logan
		MORSE, Russell
		PEPPERDINE, Burrell
		SPENCE, Ernel
		STILES, Clair
		TABOR, Donald
		TOWER, William
		WARD, George T.
		WINTERS, Daniel
1958 March 18	No. 2 Bump	WHITE, Alfred
October 23	The No. 2 Bump	ALLEN, Fidele
		AYLWARD, Ralph
		BACK, Andrew
		BOBBIE, Edward
		BOURGEOIS, Bliss
		BRINE, Henry
		BRYAN, Percy
		BURTON, Charles
		CANNING, George
		COLE, Cecil
		CORKUM, Clyde
		CROWE, Hance
		EMBREE, Harold
		EMBREE, Harry
		FRASER, Harold
		GERHARDT, Joseph
		GILLIS, Angus
		GOODE, Kenneth
		HALLIDAY, Harry
		HARRISON, Cecil
		HARRISON, Chesley
		HENWOOD, Harlan
		HOLLOWAY, Issac
		HUNTER, Hiram
		HUNTER, Wylie
		HYATT, Warren
		JACKSON, John
		JEWKES, William
		LeBLANC, Abbey

LeGERE, Alfred
LIVINGSTONE, Gilbert
MacDONALD, Arthur
MacDONALD, Edward
MacDONALD, Harold
MacFARLANE, Roy
MacKENZIE, Frank
MacKINNON, Edwin
MacLEOD, Charles
MacLEOD, Clarence
MacLEOD, Edward
MacLEOD, Frank
MacLEOD, Robert
MacLEOD, Varley
McNUTT, Harold
MADDISON, John
MARSHALL, Thomas
MILLER, Bernard
MOORING, Carl
NICHOLSON, Fred
O'BRIEN, Harry
The Bump — PERRIN, Robert
PORTER, Stirling
RAPER, Harold
RECTOR, Percy
REID, Joseph
REID, Layton
REID, Lester
REYNOLDS, Wesley
ROLFE, Ernest
ROSE, Charles
ROSS, Philip
ROSS, Robert
ROSS, St. Clair
SMITH, William
SPENCE, Percy
STEVENS, Eldon
STEVENSON, William
TABOR, Hollis
TABOR, Monty
TABOR, Raymond

			TEED, Henry TURNBULL, William WELCH, Carl WHITE, Albert WHITE, Carl
1961	May 9 at 2:35 p.m.	Syndicate Mine Slope Electrocution	VENUS, Don
1969	January 25 at 6:45 p.m.	Syndicate Mine Hit by Empty Trip on Slope	McMASTERS, Earl
	April 15 at 2:30 p.m.	Syndicate Mine Fall of Stone on Slope	GILROY, Floyd

The above list was compiled by the Cumberland Railway and Coal Company and shows a staggering 424 deaths until the closing of the mines.

January 1881 - 1891 Explosion	15
1891 EXPLOSION	125
1892 - 1956 Explosion	166
1956 EXPLOSION	39
Until the 1958 Bump	1
1958 Bump	75
1961 - 1969	3
TOTAL	424

MAYORS

Springhill became a town in 1889. Since then it has had the following mayors:

1. 1889-1891 A. E. Fraser
2. 1891-1892 William Hall
3. 1892-1894 R. H. Cooper
4. 1894-1896 William Conway
5. 1896-1898 A. A. MacKinnon
   1898-1900 William Hall
6. 1900-1909 A. B. Wilson
7. 1909-1913 E. A. Potter
8. 1913-1914 John Murray
   1914-1919 A. B. Wilson
   1919--1922 John Murray
   1922-1928 A. B. Wilson
9. 1928-1932 D. M. Smith
   1932-1934 A. B. Wilson
10. 1934-1936 W. E. Murray
    1936-1938 A. B. Wilson
11. 1938-1940 A. W. Davidson
    1940-1942 A. B. Wilson
12. 1942-1953 A. J. Mason
13. 1953-1961 R. F. Gilroy
14. 1962-1963 S. E. Wood
15. 1964-1965 R. H. MacLeod
16. 1966-1967 Arnold Jones
17. 1967- W. H. Mont

NOTES ON SOURCES

A great deal of informatiom on Springhill is contained in the Halifax newspapers: 1891 editions of the *Halifax Morning Herald,* and the *Halifax Morning Chronicle;* and the *Halifax-chronicle-herald* for the later tragedies in 1956, 1957, and 1858.

The *Springhill Record* was also a good source of information. Bertha Isabel Scott gave some facts about early history.

The reports of the two Royal Commissions that investigated the 1956 Explosion and the 1958 bump have detailed factual accounts.

Two books that deal with Springhill are R.A. Morrow's *Story of the Great Disaster at Springhill* that covers the 1891 disaster and Leonard Lerner's *Miracle at Springhill* that tells about the 1958 disaster.

In addition there are numerous magazine articles and a great many people still living who were involved in these disasters.